From Counting Numbers to Complete Ordered Fields: Set-Theoretic Construction of Real Numbers

Samuel Horelick

Copyright © 2017 by Samuel Horelick. All rights reserved.

No part of this book may be reproduced, stored in a retrieval system, or transmitted in any form or by any means, electronic, mechanical, photocopying, recording, scanning, or otherwise without prior written permission of the author, except as permitted under section 107 or 108 of the 1976 United States Copyright Act.

About the author: Dr. Samuel Horelick, PhD is mathematics professor and higher education consultant. He has graduated from three Universities with four degrees in Mathematics, Philosophy, and Mathematical Education.

CONTENTS:

Creating Integers out of Natural numbers ~ 3

Creating Rational numbers out of Integers ~ 9

Creating Real numbers out of Rational numbers ~ 14

Infinities ~ 23

Bibliography ~ 33

A science can only determine its domain of investigation up to an isomorphic mapping. In particular, it remains quite indifferent as to the 'essence' of its objects. That which distinguishes the real points in space from number triads or other interpretations of geometry one can only know by immediate intuitive perception.
Hermann Weyl

God made the whole numbers (Die ganzen Zahlen), all else is the work of man.
Leopold Kronecker

Point is that which has no parts.
Euclid

Creating Integers Out Of Natural Numbers

What is the meaning of subtraction?
What do we mean when we say "subtract x from y"?

The desire to subtract numbers from each other is inspired, for example, by an attempt to balance a checkbook; or solve simple equation such as $x + 10 = 4$.

This simple equation is not solvable if x is a natural number $x \in \{1, 2, 3,...\}$ because, *e.g.*, $x = 4 - 10 = -6$, but $x = -6$ is not a natural (counting) number. Of course, we call such negative number "integer" and accept on faith that negative numbers are just fine without thinking about where did these "negatives" come from and why is it fine to subtract 10 from 4, whereas taking $10 from a wallet which contains only $4 would be problematic.

Ancient Greek philosophers and mathematicians dismissed the idea of a negative number as absurd because the concept of negative quantity is meaningless. Even in 1803, famous French mathematician Lazaré Carnot said that to obtain a negative quantity it would be necessary to "remove something out of nothing: impossible operation…"

And indeed, this is true: if we think of "subtraction" as "removing one quantity out of another" then negative number cannot be defined – it is a logical impossibility.

If we attempt to define a negative number as a natural number multiplied by "– 1", we solve nothing because we would still have to separately define the meaning of "negative sign".

We can easily define subtraction of natural numbers if the result is also a natural number. That is, $n - m$ is a natural number only if $n > m$. It means that subtraction is only a partial binary operation on natural numbers (for all $n, m \in \mathbf{N}$, if $n < m$ then $n - m = k$ is not a natural number). But subtraction is a binary operation on the set of entities that we call "integers".

For all integers x, y, $x - y$ is an integer and every integer $k = x - y$, for some x, y.

That is, integer subtraction is a function f from $(\mathbf{Z} \times \mathbf{Z})$ to $\mathbf{Z}$ where f: $(x, y) \to k$.

It means that differences that cannot be formed in natural numbers can be formed in integers. For example, $k = 4 - 10 = -6$ is not a natural number, it is an integer.

3

So we should try to define the set of integers in terms of natural numbers in such a way as to enable us to perform subtraction (just as we have learned it in school).

For any m, n ∈ **N** with m ≥ n the difference n − m ≤ 0 exists in the set of integers. Conversely, any integer is the difference of two natural numbers n and m, such that m < n, m = n, or m > n.

An integer is a difference n − m that we can represent as a pair of natural numbers (n, m). There is a difficulty with this approach, however, because for every integer k, there are many pairs of natural numbers (n, m) such that k = n − m. For example:

10 = 30 − 20 = 12 − 2 = 73 − 63
− 7 = 14 − 21 = 43 − 50
0 = 15 − 15 = 45 − 45 = − 1 + 1

It means that subtraction cannot be used in the definition of the integers. Pairs of numbers, however, are defined in Set Theory without the use of the operation of subtraction. Cartesian product A × B is the set of all ordered pairs (a, b) such that a ∈ A and b ∈ B. Thus, we can use pairs (n, m) of natural numbers n, m to formally represent the difference n − m in the definition of integers.

Definition:

Let a, b, c, d ∈ **N**

If a − b = c − d, then both pairs (a, b) and (c, d) represent the same integer

We do not need to use subtraction in the definition of the integers, we can use addition: a − b = c − d if and only if a + d = b + c

For example:

10 − 6 = 12 − 8 iff 10 + 8 = 12 + 6

5 − 13 = 16 − 24 iff 5 + 24 = 13 + 16

22 − 22 = 46 − 46 iff 22 + 46 = 22 + 46

Therefore, we can define an integer k as a pair $(a, b) \in N \times N$ where different pairs (a, b) and $(c, d) \in N \times N$ represent the same integer if and only if $a + d = b + c$. Thus, we have defined relation between two pairs of natural numbers where $(a, b) \sim (c, d)$ iff $a + d = b + c$. This relation is an equivalence relation on the set $N \times N$ (Cartesian product set).

Proof:

Reflexivity: for all $(a, b) \in N \times N$, $(a, b) \sim (a, b)$

Let $(a, b) \in N \times N$ then $a + b = b + a$ therefore $(a, b) \sim (a, b)$.

Symmetry: for all $(a, b), (c, d) \in N \times N$, if $(a, b) \sim (c, d)$ then $(c, d) \sim (a, b)$.

Let $(a, b), (c, d) \in N \times N$
If $(a, b) \sim (c, d)$ then $a + d = b + c$ and therefore $c + b = d + a$, thus $(c, d) \sim (a, b)$.

Transitivity: for all $(a, b), (c, d), (e, f) \in N \times N$, if $(a, b) \sim (c, d)$ and $(c, d) \sim (e, f)$ then $(a, b) \sim (e, f)$

Let $(a, b), (c, d), (e, f) \in N \times N$. If $(a, b) \sim (c, d)$ and $(c, d) \sim (e, f)$ then $a + d = b + c$ and $c + f = d + e$. Then, $a + d + c + f = b + c + d + e$ and therefore, $a + f = b + e$, which means that $(a, b) \sim (e, f)$.

Therefore, relation $\sim$ is an equivalence relation that partitions the set $N \times N$ into equivalence classes. Equivalence relation $\sim$ identifies which pairs on natural numbers represent the same integer and every integer k can be represented as an equivalence class under this equivalence relation:

$$k \in Z, k = [(a, b)], \text{ where } (a, b) \in N \times N$$

This is how integers are "constructed" out of natural numbers in a logical sense of "being defined": integers are defined as equivalence classes of formal differences of natural numbers.

The set of integers Z is defined to be the set of equivalence classes $[(a, b)]$

under equivalence relation $\sim$

where $(a, b) \sim (c, d)$ where $(a, b), (c, d) \in N \times N$

if and only if $a + d = b + c$

Thus Integers are defined in terms of Natural numbers. But Integers are not just a set, it is a set together with two operations. In fact, it is one of very basic mathematical configuration.

There are two binary operations defined on the set of integers: addition and multiplication. This mathematical configuration consisting of the set of Integers and two binary operations defined on them constitutes a Ring.
Since integers are defined in terms of equivalent classes of natural numbers, the definitions of addition and multiplication are expressed in terms of equivalence classes under relation ~

If $[(a, b)]$ and $[(c, d)]$ are equivalence classes representing integers $a - b$ and $c - d$ then their sum is $(a - b) + (c - d) = (a + c) - (b + d)$, which is $[(a + c, b + d)]$.

Likewise, the product $[(a, b)] \cdot [(c, d)]$ would be $(a - b)(c - d) = ac - ad - bc + bd = (ac + bd) - (ad + bc)$, which is $[(ac + bd, ad + bc)]$.

Both the sum and the product are well-defined. Therefore, we can formally define the set of integers and their algebraic operations thus:

The set of integers Z is defined to be the set of equivalence classes $[(a, b)]$, where $(a, b), (c, d) \in N \times N$, under equivalence relation $(a, b) \sim (c, d)$ iff $a + d = b + c$

- Addition of integers is defined $[(a, b)] + [(c, d)] = [(a + c, b + d)]$

- Multiplication of integers is defined $[(a, b)] \cdot [(c, d)] = [(ac + bd, ad + bc)]$

- Additive identity is $0 = [(1, 1)]$

- For every $x = [(a, b)] \in \mathbf{Z}$, there exists $(-x) = [(b, a)]$ so that $x + (-x) = 0$

- Addition is commutative and associative

- Multiplicative identity is $1 = [(2, 1)]$

- Multiplication is commutative, associative and multiplication distributes over addition.

In view of the aforementioned, integers form a commutative ring with unity because set **Z** has the following properties:

- Additive and multiplicative associativity
- Additive and multiplicative commutativity
- Additive identity and inverse
- Multiplicative identity (no inverse in **Z**, and no 0 devisors)
- Left and right distribution

This is why we can multiply, add and subtract, with all that it entails…

John von Neumann definition of natural numbers

Thus, the set of integers is defined in terms of natural numbers. But the natural numbers themselves must be defined in terms of logically prior concepts or be accepted as a divine gift from Platonic heaven. For most of recorded history, natural numbers where understood empirically as "3 stones", "five fingers", "dozen eggs", etc. Such simplistic view identifies natural numbers with concrete numerals. In another simplistic position, numbers are identified with abstract ideas or concepts that exist independently of human mind (mathematical Platonism).

In contemporary mathematics, every concept is described by sets. All mathematical relationships are represented by membership relations between various sets. Every concept must be formally constructed from previously constructed concepts.

In the famous John von Neumann definition, natural numbers are constructed (defined) using basic set theory. Zero is defined to be the empty set $\emptyset$. Every consecutive natural number is defined in terms of all previous natural numbers. A given number N is simply the set of all von Neumann numbers less than N:

$0 = \emptyset$
$1 = \{0\} = \{\emptyset\}$
$2 = \{0, 1\} = \{\emptyset, \{\emptyset\}\}$
$3 = \{0, 1, 2\} = \{\emptyset, \{\emptyset\}, \{\{\emptyset, \{\emptyset\}\}\}\}$
...
$N = \{0, 1,..., N-1\}$

In this elegant way, all of mathematical concepts (mathematical "objects") are ultimately defined in terms of set theory. Integers are one such object. But Integers are not just a set, it is a set together with two operations. In fact, the set of Integers is one of the very basic mathematical objects.

Creating Rational numbers out of Integers: Rational numbers as equivalence classes

In this essay we will outline the construction of the set of Rational numbers Q using equivalence classes of ordered pairs (x, y) where x, y are Integers. Fractions will be formally defined as ordered pairs of Integers and Rational Numbers will be defined as equivalence classes of such ordered pairs of Integers.

Our goal is to present a formal, logically consistent definition of Rational numbers that is not based on any prior knowledge of fractions. The only prior knowledge we use is the set of Integers Z. Such formal definition is accomplished through set-theoretic construction of the set Q of equivalence classes and two binary operations defined on Q. We will see that set Q forms a commutative ring with unity where each non-zero element has multiplicative inverse. Thus, set of Rational Numbers Q is a mathematical configuration called "field". We will see that the existence of multiplicative inverse for each non-zero element of Q is precisely what enables an operation that we call "integer division". We can say that there is a need for the operation of "division" because an equation $mx = p$, where m, p are Integers, has a solution if and only if m has multiplicative inverse. That is, if there exist some operation "division" such that m "divided' by p equals to 1 (multiplicative identity).

Informally, we understand that each ordered pair (x, y) represents a fraction where x is numerator and y is denominator, but formally we treat ordered pair (x, y) as a point in two-dimensional Euclidian space E^2 where $y \neq 0$. Equivalently, we can think of ordered pair (x, y) as an element of Cartesian Product set $\{Z \times \{Z \setminus 0\}\} = \{(x, y) \mid \text{where } x \in Z \text{ and } y \in \{Z \setminus 0\}\}$.

Let's begin by defining a set Q which contains all ordered pairs (x, y) where both x, y are Integers, and y is not zero.

$$Q = \{(x, y) \mid x, y \in Z, y \neq 0\}$$

Please note that set Q is a purely abstract concept that has no relation to fractions or any other mathematical objects. The only idea we explicitly use in the construction of Q is "set membership". Implicitly, we understand what is meant by expression "set of Integers Z" because previously we have constructed set of Integers Z from the set of Natural numbers N.

The set of Natural numbers N itself was defined through set-theoretic construction (von Neumann definition) where natural numbers are constructed (defined) using basic set theory. Zero is defined to be the empty set Ø. Every consecutive natural number is defined in terms of all previous natural numbers. A given number N is simply the set of all von Neumann numbers less than N:

0 = Ø
1 = {0} = {Ø}
2 = {0, 1} = {Ø, {Ø}}
3 = {0, 1, 2} = {Ø, {Ø}, {{Ø, {Ø}}}}
...
N = {0, 1, ..., N−1}

In this way, we have constructed the set N of Natural numbers and set Z of Integers. Now, let's define a relation on the set Q where two elements (x, y) and (z, w) of Q are equivalent if and only if xw = yz

$$(x, y) \sim (z, w) \text{ iff } xw = yz$$

The operation of "multiplication" is ordinary Integer multiplication as defined in the previous presentation. This relation between the elements of Q = Z × {Z \ 0} is an equivalence relation. Therefore, this relation partitions set Q into equivalence classes.

Therefore, there is a set of equivalence classes defined by relation (x, y) ~ (z, w) if and only if xw = yz. Let's define set Q consisting of all equivalence classes under the above relation:

$$Q = \{[(x, y)] \mid (x, y) \in Q\} \text{ or } Q = \{[(x, y)] \mid x, y \in Z, y \neq 0\}$$

Equivalence relation (x, y) ~ (z, w) if and only if xw = yz identifies which pairs of Integers represent the same element of Q. Let's define two binary operations on Q:

1. Operation of Addition: [(x, y)] + [(z, w)] = [(xw + yz, yw)]
2. Operation of Multiplication: [(x, y)] · [(z, w)] = [(xz, yw)]
Note that both "addition" and "multiplication" operations are well-defined.

Let's call this mathematical configuration consisting of the set Q and two binary operations defined on Q "Rational Numbers".

Definition of Rational Numbers

Set of Rational Numbers Q is defined as a set of equivalence classes $[(x, y)]$, where $(x, y) \in Q$. Set Q is a set of equivalence classes $[(x, y)]$ under the equivalence relation $(x, y) \sim (z, w)$ iff $xw = yz$, where $x, y \in Z, y \neq 0$,

Addition in Q is defined by $[(x, y)] + [(z, w)] = [(xw + yz, yw)]$ Multiplication in Q is defined by $[(x, y)] \cdot [(z, w)] = [(xz, yw)]$

There are four special elements of Q:

- Additive Identity $[(0, 1)]$ such that $[(x, y)] + [(0, 1)] = [(x, y)]$ for all $[(x, y)] \in Q$
- Additive Inverse $[(-x, y)]$ such that $[(x, y)] + [(-x, y)] = [(0, 1)]$ for all $[(x, y)] \in Q$
- Multiplicative Identity $[(1, 1)]$ such that $[(x, y)] \cdot [(1, 1)] = [(x, y)]$ for all $[(x, y)] \in Q$
- Multiplicative Inverse $[(y, x)]$ such that $[(x, y)] \cdot [(y, x)] = [(1, 1)]$ for all $[(x, y)] \in Q$

Note: The possibility of both $x = 0$ and $y = 0$ must be excluded. If $(0, 0)$ is an element of Q then $(0, 0)$ is related to all elements (c, d) of Q because $0(d) = 0(c)$, for all $c, d \in Z$.

If one element of a set is related to all elements of this set then either
a. The relation is not transitive, or
b. Transitivity would imply that all elements are related to each other.

This is how Rational numbers are "constructed out of" Integers in a logical sense of "being defined" in terms of a previously constructed set. Rational numbers are defined as equivalence classes of ordered pairs (x, y), $y \neq 0$ under the equivalence relation $(x, y) \sim (z, w)$ iff $xw = yz$.

The set of Rational Numbers $Q = \{[(x, y)] \mid x, y \in Z, y \neq 0\}$

In view of the aforementioned, Rational Numbers form a Commutative Ring with unity because Q has the following properties:

- Additive and Multiplicative Associativity
- Additive and Multiplicative Commutativity
- Additive Identity and Inverse
- Multiplicative Identity
- Multiplicative Inverse (for every non-zero element)
- Left and Right Distribution

Note: Commutative Ring with unity where every non-zero element is invertable is called a "Field". Set of Rational numbers is not just a collection of elements; it is a structure where elements may be combined in specific ways under specific rules (such as being able to solve equations of the form $mx = p$, where $m, p \in Z$).

Number Sets and the Question of Inclusion

As was presented previously in the famous John von Neumann definition, Natural numbers are constructed (defined) using basic set theory. Zero is defined to be the empty set $\emptyset$. Every consecutive natural number is defined in terms of all previous natural numbers.

A given number N is simply the set of all von Neumann numbers less than N:

$0 = \emptyset$
$1 = \{0\} = \{\emptyset\}$
$2 = \{0, 1\} = \{\emptyset, \{\emptyset\}\}$
$3 = \{0, 1, 2\} = \{\emptyset, \{\emptyset\}, \{\{\emptyset, \{\emptyset\}\}\}\}$
...
$N = \{0, 1, ..., N-1\}$

Informally, we say that set Q contains set Z and set Z contains set N. But that is informally. Formally, we have to say that set Z contains the image of N under some mapping and set Q contains the image of Z (and consequently Q contains the image of N) under some other mapping.

It must be emphasized that sets N, Z, Q are completely different sets. Set itself and its image under some function (mapping) are two different mathematical entities, set-theoretically speaking.

Let's define a function from the set of Natural numbers to the set of Integers:

$f: N \to Z$, $f(n) = [(n, 0)]$

As presented previously, set of integers Z is defined to be the set of equivalence classes $[(a, b)]$ under equivalence relation $(a, b) \sim (c, d)$ if and only if $a + d = b + c$, $(a, b), (c, d) \in N \times N$.

Note that $(a + d) = (b + c)$ if and only if $(a - b) = (c - d)$.

This function maps every Natural number n onto an equivalence class [(n, 0)]. Thus, there is a "perfect copy" of N "sitting inside" Z. Moreover, operations of addition, multiplication, and the ordering of N are preserved in Z. It can be shown that such function f is unique. Thus, function f : N → Z , f(n) = [(n, 0)] is an isomorphism and "canonical embedding".

Since we have set N, a "perfect copy" of N that is "included" in Z, and a specific mapping identifying the copy with the original set N, we can say that set of Natural numbers is a subset of Integers.

We can say that by identifying set N with its copy because there is no difference between them: all operations in one set are identical to the same operations in the other set. But, set-theoretically speaking, set N and its image in Z are two different set, two different mathematical objects.

In the same way, sets Q and Z are completely different sets. There is an injective function from the set of Integers to the set of Rational numbers g: Z → Q defined by g(k) = [(k, 1)] where [(k, 1)] is an equivalence class as defined above. This function preserves all operations and order, it is an isomorphism, it's unique, and so it is a "canonical embedding". Again, we have a perfect copy of Z included in Q and a specific mapping identifying the copy with the original. But again, sets N, Z, and Q are different mathematical objects.

Furthermore, we can do the same thing and show that the set Rational numbers is included in the set of Real numbers by identifying members of Q with specific Dedekind cuts or specific equivalence classes of Cauchy sequences. We can construct a unique, injective mapping from Q to R that will preserve all operations and so we would be able to say that the set of Rational numbers is included in the set of Real numbers (and by extension, sets N and Z are also included in the set of Real Numbers).

In the same way, it can be shown that set of Real Numbers is included in the set of Complex numbers C and therefore N, Z, Q are also included in C. This gives us a well-known "fact": N ⊆ Z ⊆ Q ⊆ R ⊆ C.

We just have to remember that each set was independently constructed and therefore N, Z, Q, R, C are completely different sets; completely different mathematical objects.

Creating Real Numbers out of Rational numbers:
Set-theoretic Construction of Number Sets

We can see that that inability to solve some simple equations within the existing number sets has led to the development of "new" number sets. For example, if all we have is the set of Natural numbers $\{1, 2, 3, …\}$ then we cannot solve any equations of the form $x + a = 0$. We might clearly intuit that $x = -a$, but "negative" number is not a meaningful concept within the set of Natural ("counting") numbers. Therefore, a new number set must be defined. The one that allows the existence of a "negative" numbers a so that $x + a = 0$, for all Natural numbers x.

- Equation $x + a = 0$, $x \in \mathbb{N}$ is not solvable in natural numbers $\mathbb{N}$.

Thus, the set of Integers is constructed from Natural numbers. The integers Z are defined to be the set of equivalence classes ("formal differences") of ordered pairs of natural numbers (see Creating Integers Out Of Natural Numbers).

Similarly, we cannot solve any equations of the form $ax + b = 0$ in Integers. We might clearly intuit that $x = -b/a$, but "division" of two Integers is not a meaningful concept within the set of Integers. Therefore, a new number set must be defined. The one that defines the meaning of "division of two integers".

- Equation $ax + b = 0$, $x \in \mathbb{Z}$, $a \neq 0$ is not solvable in integers $\mathbb{Z}$.

Thus, the set of Rational numbers is constructed from Integers. Rational numbers are defined to be the set of equivalence classes ("formal quotients") of ordered pairs of integers (see Creating Rational Numbers Out of Integers).

Not all quantities, however, can be represented by Natural, Integer, or Rational numbers. For example, equation $x^2 - 2 = 0$ implies that $x = \sqrt{2}$. But it turns out that $\sqrt{2}$ cannot be represented in the form n/m, where n, m are integers.

- Equation $a^2x + bx + c = 0$ a, b, c $\in \mathbb{Q}$ is not solvable in rational numbers $\mathbb{Q}$. For example, if $a = 1$, $b = -2$, $c = 0$ then $x^2 = 2$, but $x = \sqrt{2}$ is not in $\mathbb{Q}$.

The number $\sqrt{2}$ cannot be represented in the form n/m, where n, m are integers. Therefore, $\sqrt{2}$ is not a Rational number. In this essay, we will discuss an axiomatic constructions of Real numbers in terms of Rational (and irrational) numbers.

Discovery of Irrational Numbers (5th century BC)

Intuitively obvious idea: given any two line segments A and B, there is a third line segment C that can be "marked off" a whole number of times into both A and B. That is, every magnitude (length) is expressible as a ratio p/q of two integers p, q. In particular, it was believed that the length of a diagonal of a square can be expressed as a rational number p/q, just as any length.

It went like this: consider a square with side S and diagonal d. If there exist a third segment t such that t can be marked off a whole number of times into S and d then S = qt and d = pt for some integers p, q. But d = S$\sqrt{2}$, therefore, pt = (qt)$\sqrt{2}$ and so (p/q) = $\sqrt{2}$, which is a rational number. (Note: $S^2 + S^2 = d^2$, $2S^2 = d^2$, thus d = S$\sqrt{2}$).

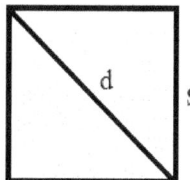

Intuitively Obvious Idea: p/q = $\sqrt{2}$ is a rational number.

Pythagoreans, a religious and philosophical school founded by Pythagoras of Samos (569BC to 475BC) in Southern Italy, made very significant contributions to mathematics including a famous discovery that not all numbers are rational.

Classic disproof of an *Intuitively Obvious Idea* – a proof by contradiction that $\sqrt{2}$ is not a rational number:

- Assume that integers p and q have no common factor other than unity.
- Let p/q =$\sqrt{2}$ → $p^2/q^2 = 2$ → $p^2 = 2q^2$ → p is even
- Let p = 2r for some integer r
- Then $p^2/q^2 = 2$ → $p^2 = (2r)^2 = 4r^2 = 2q^2$ or $2r^2 = q^2$ which means that q is even, *which contradicts the assumption that p and q have no common factors.*
- Thus, there are no integers p, q such that p/q = $\sqrt{2}$.
- Therefore, $\sqrt{2}$ is not a rational number.

Intuitive Approach: Visualizing Continuous Number Line

Rational number line is not continuous: it has gaps (non-rational numbers). Rational numbers are "densely ordered" – between any two Rational numbers are other rational numbers, but also irrational numbers. Geometric line is continuous. The points on a line are dense, meaning that real line number has no gaps - between any two points on the real line is an infinity of other points. Therefore, any definition of a Real number must produce a continuous Real number line.

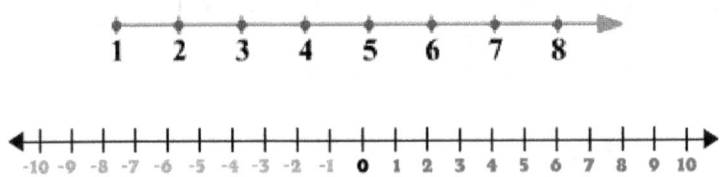

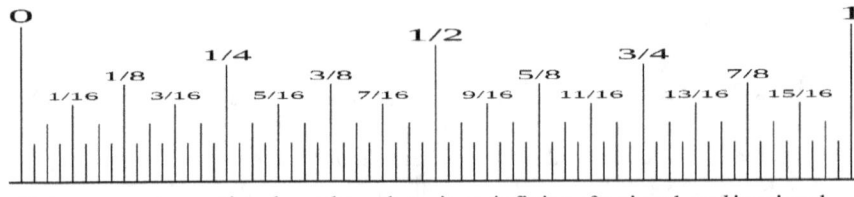

Between any two rational numbers there is an infinity of rational *and* irrational numbers

We need to connect our intuitive sense of a continuous number line with a detailed account of how Real numbers are logically (axiomatically) constructed "out of" Rational numbers. Rational number line is not continuous – it has "holes" for non-rational numbers such as $\sqrt{2}$, π, e, and infinitely many other Irrational numbers.

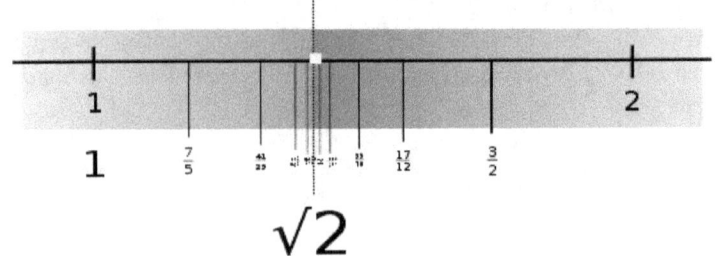

Real number line **is** continuous: between any two points (real numbers) on the real line there is an infinity of other points (real numbers). The points on the real line are dense, meaning that the real number line has no gaps

Dedekind cut – formalization of the idea of continuity.

J. Richard Dedekind created first rigorous definition of Real number around 1860. In his work on the foundations of mathematical analysis, he returned to the old Greek definition of the equivalence of two ratios (Eudoxus of Cnidus, ca. 360 BC, about 2375 years ago).
That definition characterized each rational number a/b by two sets of rational numbers, those less than a/b and those greater than a/b.

In his own words: "I felt … the lack of a really scientific foundation for arithmetic in discussing the notion of the approach of a variable magnitude to a fixed limiting value … This feeling of dissatisfaction was so overpowering that I made the fixed resolve to keep mediating on the question till I should find a purely arithmetic and perfectly rigorous foundation for the principles of infinitesimal analysis" (*Essays on the Theory of Numbers: Continuity and irrational numbers*, 1872).

Given any rational number, all other numbers (points) on a straight line are divided into two classes A and B such that every point in A is to the left of (less than) every point in B.

Every gap in rational numbers also divides Rational number line into two sets – one to the left and one to the right of a gap.
Consider Set $L = \{x, r \in \mathbb{Q} \mid x < r\}$ that is not empty, that has no largest element and, if $x \in L$ then so is any rational number less than x (set L is called "down set").

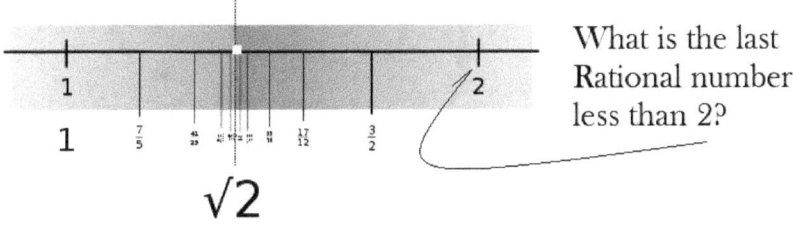

What is the last Rational number less than 2?

17

Subset L of $\mathbb{Q}$ is called "Dedekind cut" in the Rational numbers if it satisfies:

1. L is not empty and $L \neq \mathbb{Q}$
2. L has no largest element
3. If $x \in L$ then so is any rational number less than x

Real number is defined to be that gap, called "Dedekind cut", that separates Rational number line into two classes of numbers. We can think of each number x corresponding to Dedekind cut as a set of rational numbers to the left of (less than) number x. Each rational number r determines a cut, *but there are Dedekind cuts that do not correspond to any rational number*.

For example, set $B = \{x \in Q \mid x < \sqrt{2}\}$ is certainly Dedekind cut because it satisfies all three conditions but $r = \sqrt{2}$ is not a rational number. So, Dedekind cuts come in two flavors – one made by rational number and another made by "irrational" number.

Dedekind cut in the set of Rational numbers (L|R):

1. Set L has a largest element and R has no smallest element.

Example: L is a set of all rational numbers less than or equal to 1:
$L = \{x \in Q, x \leq 1\}$, $R = \{x \in Q, x > 1\}$

2. Set L has no largest element and set R has smallest element.

Example: L is a set of all rational numbers less than 1.
$A = \{x \in Q, x < 1\}$, $R = \{x \in Q, x \geq 1\}$

3. Set L has no largest element and set R has no smallest element.

Example:

L is a set of all rational numbers whose square is less than 2:
$L = \{x \in Q, 0 \leq x^2 \leq 2\}$, $R = \{x \in Q, x^2 > 2\}$

Dedekind cut of the type 1 or 2 is called "rational cut" or "rational number":

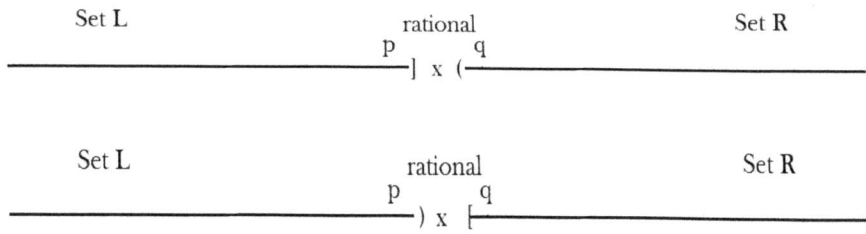

Dedekind cut of type 3 is called "irrational cut" or "irrational number":

```
        Set L                                              Set R
                              p           q
    ————————————————————————— ) y (————————————————————————
                              irrational
```

The set of Real numbers is a union of all rational cuts and irrational cuts.

Note: There are many more irrational numbers that rational numbers (in a special sense of "more" that is applicable to infinite sets). As a consequence of Cantor's proof that the Real numbers are uncountable and the Rational numbers are countable, it follows that Irrational numbers are uncountably infinite (there can be no one-to-one correspondence between natural numbers and irrational numbers).

<u>Real numbers form Complete Ordered Field</u>

I. Configuration $(R, +, *)$ is an algebraic field because it satisfies the following:

1. Additive and Multiplicative Associativity
2. Additive and Multiplicative Commutativity
3. Additive Identity and Inverse
4. Multiplicative Identity
5. Multiplicative Inverse (for every non-zero element)
6. Left and Right Distribution

Note: Set of Real numbers is not just a collection of elements (numbers).
The set of Real numbers is a structure where elements may be combined in specific ways under specific rules. One of the rules, for example, is being able to solve equations of the form $x + a = 0$, $ax + b = 0$, $a^2x + bx + c = 0$.

II. Algebraic field (R, +, *) is totally ordered because it satisfies the following:

1. For any elements x, y, z ∈ R, either a ≤ b or b ≤ a (comparability criterion)
2. Reflexive: a ≤ a
3. Anti-symmetric: if a ≥ b and a ≤ b then a = b
4. Transitive: if a ≤ b and b ≤ c then a ≤ c

One special property distinguishes Real numbers system from Rational number system. Without this property, Calculus would not be possible - most theorems would be false. This property is "Completeness". A totally ordered set is complete if every nonempty subset that has an upper bound, has a least upper bound. For example, the set of real numbers R is complete but the set of rational numbers Q is not.

A subset M of an ordered field F is said to be bounded above if there is an element m ∈ F such that for every x ∈ S, x ≤ m. The element m is called least upper bound if m is less than every other upper bound of subset M⊂F.

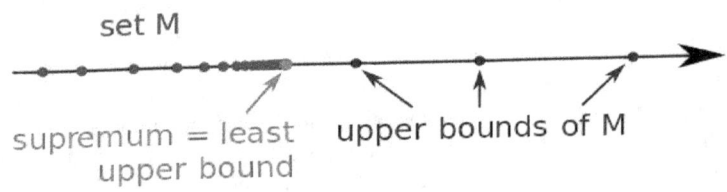

III. Real Number Field is Complete

Any non-empty subset of R that is bounded above has a least upper bound (not necessary in the subset itself). Thus, Real numbers form Complete Ordered Field. Equivalently, Real numbers are an ordered field of Dedekind cuts. Note: all subsets of Rational numbers are also bounded above, but some of them do not have a least upper bound in the set of Rational numbers.

IV. If every non-empty subset of a field F which is bounded above has a least upper bound then no cut in A is a gap.

V. Complete Ordered Field of Real Numbers is Unique

If (F, +, *) is a Complete Ordered Field then F is isomorphic to R.

There is only one complete ordered field in the sense that any two complete ordered fields are isomorphic. Isomorphism is a bijective mapping between two mathematical configurations that preserves sets and relations among elements of these sets. Two configurations are isomorphic if they can be proven to have exactly the same mathematical structure. That is, Real numbers are in some ways a "constant of nature".

Addendum:

1. Intuitively, a space is complete if there are no "points missing" from it. For instance, the set of rational numbers is not complete, because $\sqrt{2}$ is "missing" from it, even though one can construct a Cauchy sequence of rational numbers that converges to it:

2. The Least Upper Bound property "fills the holes" in Rational number line. Real numbers are "complete" forming a Complete Ordered Field in contrast to Rational numbers which only form Ordered Field.

3. Geometrically speaking, some lengths of line segments, such hypotenuse of right triangle with sides equal to 1 unit, are not rational numbers. However, length of any line segment can be associated with a real number (Cantor-Dedekind Axiom: real numbers are order-isomorphic to the linear continuum. That is, there is one-to-one correspondence between real numbers and points on a line.).

4. In mathematical analysis, a metric space M is called complete (or a Cauchy space) if every Cauchy sequence of points in M has a limit that is also in M or, alternatively, if every Cauchy sequence in M converges in M.

5. Just after Napoleonic wars, Augustin Cauchy developed rigorous definition of a limit and convergent sequence. In 1830's Bernard Bolzano attempted to develop a theory of real numbers as limits of rational number sequences. Charles Méray pointed out in 1869 that defining a limit as a real number and then, in turn, defining real number as a limit of a sequence of rational numbers commits a fallacy of *petition principii* (begging a question or vicious circle).

Karl Weierstrass (Bolzano-Weierstrass Theorem), Eduard Heine, Georg Cantor (Heine-Cantor Development) have contributed to the definition of real number as convergent sequences of rational numbers. Those sequences that fail to converge to a rational number, converge to "irrational" number, instead.

6. Recursive definition of a sequence: $a_{n+1} = (a_n + 2/a_n)/2$, where $a_1 = 1$:

$1, {}^3/_2, {}^{17}/_{12}, {}^{577}/_{408}, {}^{665857}/_{470832}, \ldots \cong 1, 1.5, 1.41667, 1.414215, 1{,}414213562, \ldots$

The limit of this sequence is $\sqrt{2}$.

Infinities

How to decide if two sets are of the same size without the appeal to a concept of "size"?

Well, if we could find a way to assign an element of one set to an element of the other set so that each element of one set is assigned to exactly one, unique element of another set and no elements are left over unassigned then these two sets are of the same size. That is, if we can show a one-to-one and onto (*i.e.,* bijective) function between the sets then we know that these two sets have exactly the same number of elements. The concept of being "of the same size" is quite irrelevant here. This method is obviously true for any given finite set of things. Such a finite set can be put into one-to-one correspondence with some subset of integers, say, from 1 to some integer n. This number n is called *cardinality* of the set.

If the set is infinite, we can attempt to put it into one-to-one correspondence with the set all of integers. If we can show some methodical and well-defined way of doing it (that is, if can show a bijective function between N and a given set A) then we call this set A *countable* (because we can count it, literally, even if it will take eternity). For example, the set of all even numbers is infinite, but we can define a function f on N such $f(n) = 2n$. It is a bijective function and so the set of all even integers is of the same size (*cardinality*) as the set of all integers – countably infinite.

Set is called "finite" if it can be put into one-to-one correspondence with the set of Natural numbers $\{1, 2, 3, ..., n\}$. The original definition of "infinite set" was simply this: if there is no such natural number n then set A is infinite. If we remove n elements from an infinite set, we would have some elements left over, in fact, infinitely many of them. It seems to contradict basic and intuitively obvious principle that whole is bigger than its part. So a question of the meaning of "size" still comes up. The formal definition is: set A is smaller in size than set B if

i). A can be put in one-to-one correspondence with a proper subset of B, and
ii). A cannot be put into one-to-one correspondence with all of B.

Or, more elegantly, there is an injection from A to B but no bijection.

This fact is elegantly illustrated by famous "Hilbert hotel" example. A hotel with just so many rooms can accommodate only so many guests if each guest stays in one room an each room has only one guest.

An infinite hotel can accommodate infinitely many guests. Let's say 5 guests arrive to find an infinite hotel fully booked. Can they still get in, one per room? Of course, the administration can simply ask each guest to move 5 rooms down the hall and, presto, you have 5 rooms to accommodate 5 new guests.

Now, imagine that infinity of new guests arrives. Can they all get in? One solution would be to ask each old guest to move to a new room with a room number double of the old room number, *e.g.*, 1 to 2, 5 to 10, 50 to 100. That is, we define a function f on N such $f(n) = 2n$. And so infinity of guests moves in, no problem. The infinity is, of course, a countable infinity.

Geometric demonstration that an infinite set is of "the same size" as its subset. Equivalently: points on a line are dense; or, line is infinitely divisible; or, a point has no magnitude.

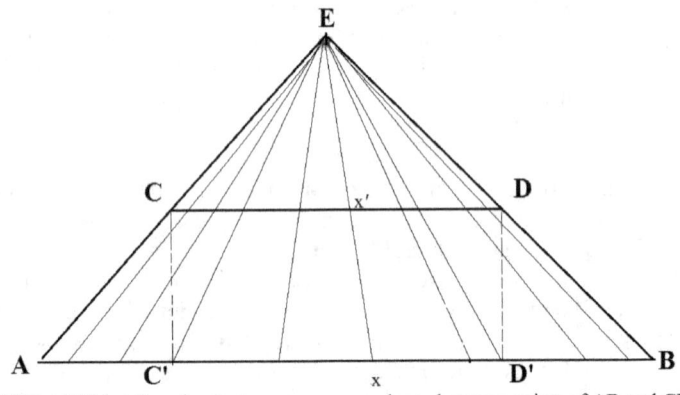

|CD| < |AB| but there is one-to-one correspondence between points of AB and CD

Every line from point E to the base of triangle AB intersects the line CD as well as the line AB. Therefore, there is one-to-one correspondence between each point of CD and each point of AB. The length of CD, however, is obviously less than the length of AB.

Now, the question would be "are all infinities of the same size?" Meaning, are all infinities countable? For example, the set of Rational numbers is dense, meaning that between any two rational numbers there are infinitely many other rationals. This suggests that set Q should be in some sense "bigger" than set N. Well, it is another instance when intuition is not a good guide to truth. Let's arrange all rational numbers in this manner (any arrangement would do, actually):

1/1 1/2 1/3 1/4 ...
2/1 2/2 2/3 2/4 ...
3/1 3/2 3/3 3/4 ...
4/1 4/2 4/3 4/4 ...
.
.

Such table contains all rational numbers. And we can definitely count them, even though there are infinitely many duplications as 1/1, 2/2, 3/3, 4/4, etc. Thus, Rationals are countable, somewhat counter-intuitively. We can also think of a fraction a/b as an ordered pair (a, b) and so the set of fractions of the form ka/kb forms an equivalence class of a/b, for any natural (integer) k.

The Real numbers, though, are not countable (that is, *uncountable*). For a proof, let's assume that Reals are countable and so we should be able to arrange all Reals into one-to-one correspondence with Natural numbers. Well, let's draw a picture of such one-to-one correspondence between sets N and R:

1 – 2.37985567...
2 – 7.75687475...
3 – 3.97887989...
4 – 0.67846893...
.
.

This array represents all naturals and all reals. If we could find at least one real number that cannot possibly be in this array, then Reals would not be countable. Let's pick any integer for the integer part of a decimal expansion of a real number. Let's choose the first decimal digit to be different from the first decimal digit of the first number in our array. And the second decimal digit should be different from the second decimal digit of the second number in our array, and so on, *ad infinitum*. So, no matter how huge our array is, there will always be a real number not in it. Thus, there is no one-to-one correspondence with N and, therefore, R is uncountable. It is "larger" than N, colloquially speaking. (This is a famous Cantor's Diagonal Argument).

Now, having countable and uncountable infinities, we can reasonably ask if the set of all subsets of Natural numbers is countable. Well, the set of all finite subsets of N is certainly countable. But *the set of all sets* includes, well, *all sets*, finite and infinite (including N itself).

Such set of all sets is called *power set*. Power set of N, denoted P(N) consists of all subsets of N. Again, let's assume that P(N) is countable. First, let set X = {a, b, c, d, e, ...} be the largest infinite set. The power set of X will contain all subsets of X, P(X) = {{a}, {a, b}, (b, c, e}, ...}. Now, P(X) is itself a set so by our assumption it cannot be larger than X. Obviously, it cannot be smaller because it contains all elements of X such as {a}, {b}, {c}, etc. Therefore, cardinality of X and P(X) must be the same – there must be one-to-one correspondence between X and P(X).

In particular, we can match all elements of X with elements of P(X) that contain them. For example, element *a* with set {*a*, b, c}, *d* with (p, k, *d*, w}, etc. But, if there is one-to-one correspondence between X and P(X), there will be elements of X that are assigned to elements of P(X) that do not contain them, for example element *m* with the set {s, f, z}. Consider a set of all elements of X that are *not* matched with elements of P(X) that contain them. Let's call it set F.

This set F is itself a subset of X, so *it must have been matched with some member of set X*. What could this member be? After all, set F was selected of precisely those elements of X that are not matched. This is a formal contradiction and, therefore, no match-up is possible. Set P(X) is *cardinally* larger than set X.

This brilliant argument is a proof of Cantor's theorem: For any set A, the power set of A is larger than A. (That is, cardinality of P(A) is larger than cardinality of A).

So, as they used to say in novels, "the plot thickens". For we can ask, is there no largest cardinality then? And the answer is "certainly", there is no largest cardinality because for any given set, *e.g.,* power set of B, we can get power set of the power set of B: P(P(B)) and its cardinality is sure way bigger than cardinality of P(B), not to mention B itself. So, the sizes of sets are endless.

Unfortunately, this clear and perfectly logical proof leads to an obvious contradiction. Namely, let's take a "universal set" – the set of all sets. Surely, it is bigger than any other set for it contains them all. However, since we defined a concept of a power set, we can form a power set of this universal set. And the cardinality of this power set is surely larger than the universal set. (Cantor's Paradox, 1897).

This is a sad conclusion because it demonstrates that there is something very wrong with our (human) thinking process.

After all, if we can think of a set of objects, we can think of universal set containing all objects. Moreover, if we can think of some property, we can think of a set of objects having that property. If a property is being a college teacher, we can think of a set of all college teachers. For a property of being divisible by two, we have a set of all even numbers. No problem. But when we generalize to the set of all sets, we get Cantor's paradox, which is insoluble in "naïve" set theory.

An equivalent (but more famous) paradox was constructed by Russell. The essence of it is that our logic leads to a contradiction, to a statement p and not-p.

Consider a set of all teachers. Clearly, this set is not a teacher. But a set of all abstract ideas is itself an abstract idea. So we can say that all sets can be divided in two classes – those that are members of themselves (*e.g.,* abstract ideas) and those that are not (*e.g.,* teachers).

Let's call sets are not members of themselves ordinary sets and let's call those sets that are members of themselves extraordinary sets.

Let B be a set of *all* ordinary sets. Is B itself an ordinary set, *i.e.,* is B not a member of itself or is it extraordinary, *i.e.,* member of itself?
Suppose that B is an ordinary set. All ordinary sets belong to B and so B belongs to B.

But, if B belongs to B then it is extraordinary by definition. Likewise, if B is extraordinary then it is a member of itself. But by definition B contains only those sets that are ordinary. Thus, a formal contradiction.

What is a contradiction? It's when you start with some well-defined propositions (premises), then you carefully follow logical inferences, and then you end up with a statement p *and* not-p (a formal contradiction).

After such traumatic experience, it may be wise to reconsider the *clarity* of your propositions and, possibly, the *validity* of your logical inference. And that is precisely what was done.

It seems obvious that for any property, we can define a set of things having this property. But it leads to a fatal contradiction – Cantor's and Russell's paradoxes (and many similar paradoxes).

The fallacy here is to affirm that there exists such a thing as a set of all sets or a set of all ordinary sets. It is not logical inference that's at fault. It's the starting point that's fallacious.

The principle that for any property, we can define a set of things having this property is called *unlimited abstraction principle*. It is due to one of the greatest mathematicians and logicians Gottlob Frege. He created a system of axioms that, seemingly, all of mathematics could be derived from just a few propositions. Unfortunately, he was proved blatantly incorrect.

This problem was corrected by substituting <u>*limited* abstraction (separation) principle</u> instead of <u>*unlimited*</u> abstraction principle (among other things). This new principle states that given any set A and any property defined on A, you can construct a set of elements of A having that property.

It does not sound revolutionary, but it most certainly is. This was done by another great light – Ernst Zermelo.

Zermelo axioms of set theory:

1. Separation principle
2. Existence of empty set
3. For any sets X, Y, there exist set {X, Y} whose members are just X and Y
4. For any set A, there exist P(A)
5. For any set A, there exist its union
6. There exists the set of Natural numbers (axiom of infinity).

Later work by Abraham Fraenkel and others resulted in the addition of Axiom of regularity, Axiom of replacement, and Axiom of choice. This new system, called Zermeo-Fraenkel with Axiom of Choice (ZFC), is considered to be the basis of all modern axiomatisations of set theory and, therefore, for all contemporary mathematics. However, there are some serious challenges against it.

We must now take a short detour so as to look at the problems from a proper perspective.

Essentials of Axiomatic method

In a valid argument, premises imply conclusion:

(p_1 and p_2 and p_3 and ... and p_n) → q

But what if a premise p_i is itself being implied by some other, logically prior, statement x? To avoid a chain of pointless circular definitions, we must *assume* some statements as so basic as not to require an explicit definition in terms of *something else*.

The "axioms" are such primary statements of logical discourse – every true statement must be derivable form the axioms and, possibly, other previously derived statements via the rules of logical inference. When we say that statement X is proved in an axiom system Y, we mean that X was derived from the axioms of Y (and, perhaps, some other statement A, B, C that were derived from the axioms of the system Y previously to X).

For example, all of Euclidian geometry was derived from five postulates (axioms):

1. There exists a line between two points
2. For every segment, there exists a unique mid-point
3. For every point, there exists a circle with a given radius
4. All right angles are congruent to each other
5. Parallel lines do not intersect if continued to infinity (This postulate is independent from the other four.)

Every geometric theorem is derived from these postulates (and perhaps other logically prior true statements that were derived from the postulates earlier).

Cantor's theorem

For any set A, the power set of A is larger than A. [*Note:* For any set A, the set of all subsets of A is called "Power set of A" denoted $P(A)$]. That is, cardinality of $P(A)$ is larger than cardinality of A. In other words, there is an injection from A to $P(A)$ but no bijection.

There is no largest cardinality because for any given set, *e.g.*, power set of B, we can get power set of the power set of B, *i.e.*, $P(P(B))$ and its cardinality is bigger than cardinality of $P(B)$, not to mention B itself. So, the sizes of sets are endless.

Unfortunately, this clear and perfectly logical proof leads to a contradiction. Namely, let's take a "universal set" U – the set of all sets. Surely, it is bigger than any other set for it contains them all. However, since we defined a concept of a power set, we can form a power set P (U) of this universal set. And the cardinality of this power set is surely larger than the universal set. (Cantor's Paradox, 1897).

Unlimited Abstraction Principle

It seems obvious that for any property we can define a set of things having this property. If a property is being a college teacher, we can think of a set of all college teachers. For a property of being divisible by two, we have a set of all even numbers.

But when we generalize to the set of all sets, we get a formal contradiction (Cantor's paradox). The fallacy here is to affirm that there exists such a thing as a set of all sets.

The principle that for any property we can define a set of things having this property is called *unlimited abstraction principle*. It is due to one of the greatest mathematicians and logicians Gottlob Frege. He created a system of axioms whereby all of mathematics could be derived from just a few propositions. Unfortunately, this system leads to a contradiction (statement "p and not-p").

Limited Abstraction Principle

This problem was corrected by substituting *limited abstraction (separation) principle instead of unlimited abstraction principle* (among other things). This new principle states that given any set A and any property defined on A, you can construct a set of elements of A having that property. It does not sound revolutionary, but it most certainly is. It was done by another great light – Ernst Zermelo. Whereas all of Set Theory could be derived from Frege's Axioms of logic and the unlimited abstraction principle, this is not the case with Zermelo system. Zermelo had to explicitly postulate the following axioms (axioms of existence):

1. Separation principle
2. Existence of empty set
3. For any sets X, Y, there exist set {X,Y} whose members are just X and Y
4. For any set A, there exist P (A)
5. For any set A, there exist its union
6. There exists the set of Natural numbers (axiom of infinity)

Continuum Hypotheses

Since cardinality of $P(A)$ is larger than cardinality of A, for any set A (finite or infinite), is there a set with cardinality intermediate between A and $P(A)$?

Cantor conjectured that there is no such set. This conjecture is called Continuum *hypotheses* because it has not been proven and it has not been disproven. It is called "Continuum" because $P(N)$ can be put into one-to-one correspondence with the set of Real numbers ("continuum").

Undecidability of Continuum Hypotheses in Zermelo–Fraenkel system

Kurt Gödel proved in 1939 that CH *cannot be disproved* in ZF axiom system.
In 1963, Paul Cohen *proved that CH cannot be proved* in ZF, either.
Therefore, Continuum Hypotheses is *independent* of the axioms of ZF.
That is, it is "undecidable" in ZF.

Note that the question here is not whether CH is "true" or "false" in some sense. The question is whether CH or its negation can be proved in ZF, that is, if CH can be *derived from the axioms* of ZF.

And that would take us into Mathematical Philosophy – the schools of mathematical "realists", "formalists", and "constructivists", and beyond.
And that is far too far for such modest manuscript.

Note on objections to Zermelo system:

Many greatest lights (e.g., Gauss, Poincare, Kronecker) objected to the whole idea that infinity is actually a well-defined concept, a mathematical entity whose "size" can be defined and compared. If *these* ideas are true then Zermelo solved nothing. At best, he simply built one internally consistent metaphysical theory instead of another.

Also, Zermelo axioms for set theory have not been proven consistent. We can only say that we are 'reasonably' sure they are not contradictory, which is to say precisely nothing. There were and are very serious objections to the axioms, the chief being that they are arbitrary (*ad hoc* solution, so to speak), chosen to fit a situation with Frege's System (contradictions) and so enable an effective program of basing math on set theory.

But Zermelo axioms are not based on any previously established truths (obviously). In this case, to show that Z-system is a legitimate basis for further work, the consistency of Z-system must be formally demonstrated. Otherwise, the axioms are no more than really clever definitions of some fundamental notions, whose consistency must be *assumed*.

Needless to say, Z-system's consistency can only be proven in another system of previously (and independently) proven consistency, of which Z-system is but a part. If such system existed (or could ever be constructed) then it could be the basis for all mathematics, regardless of the usefulness of Z-system.

That is to say, if we *could* prove the consistency of Zermelo Axioms of Set Theory then we *would not need* Zermelo-Fraenkel system as a basis to establish the whole superstructure of mathematics. Because then we could base all mathematics on that "super–system" in which we proved the consistency of Z-system in the first place…

Bibliography

Essays on the Theory of Numbers, R. Dedekind
The Structure of The Real Number System, L. Cohen and G. Ehrlich
The Real Numbers: An Introduction to Set Theory and Analysis, J. Stillwell
Fundamentals of Abstract Analysis, A. Gleason
Foundations and Fundamental Concepts of Mathematics, Howard Eves
The Philosophy of Set Theory, M. Tiles

www.ingramcontent.com/pod-product-compliance
Lightning Source LLC
Chambersburg PA
CBHW050035230526
45470CB00003B/1290